George B. Brainard

The Water Works of Brooklyn

A historical and descriptive Account of the Construction of the Works

George B. Brainard

The Water Works of Brooklyn
A historical and descriptive Account of the Construction of the Works

ISBN/EAN: 9783337139506

Printed in Europe, USA, Canada, Australia, Japan

Cover: Foto ©ninafisch / pixelio.de

More available books at **www.hansebooks.com**

THE

WATER WORKS

BROOKLYN

A HISTORICAL AND DESCRIPTIVE ACCOUNT
OF THE CONSTRUCTION OF THE
WORKS, AND THE QUANTITY,
QUALITY AND COST OF
THE SUPPLY.

BROOKLYN:
1873.

PREFACE.

An abundant supply of pure and wholesome water is one of the first requirements of a growing city. The applicances necessary to procure this supply usually constitute one of the most important and expensive branches of the Public Works. The progress of an undertaking of this character is duly chronicled in the daily prints, and at its completion the citizens are generally familiar with the details; but as time goes on, the population grows, receiving additions from abroad, and the question soon is asked with increasing frequency: "Whence does the water come, and how is it conveyed?" Since the introduction of the water into the city of Brooklyn, there has been no description of the Water Works accessible to the general public, with the exception of the costly official memoir and, the scattering articles in the daily papers. To supply, in a condensed and inexpensive form, the result of a compilation from all existing sources of information, together with facts gained by several years of personal connection with the Works, as an answer to all questions that may be asked concerning the Ridgewood Water, is the present design of the

AUTHOR.

Brooklyn Water-Works.

History.

THE first agitation of the question of Water-Supply of which we have authentic record occurred in 1834, shortly after the incorporation of Brooklyn as a city, and when it contained but 23,000 inhabitants. It was then proposed to sink wells near the foot of Fort Green, from which the water was to be forced by steam pumps to a reservoir upon the summit of the hill. The distribution was to consist of 11 miles of 10 and 4-inch pipe; the whole to cost $100,000. This proposition was not, however, acted upon. In 1847 the subject was again brought up by a committee-report, advising the construction of immense wells at the foot of the southeast slope of the hills at the rear of the city; the water to be raised by steam-power to a reservoir 40 feet above the tops of the houses on the Heights.

Again in 1849 the well-system was advocated, but received a severe blow through the report of Dr. Torrey that the Brooklyn well-water contained $18\frac{1}{4}$ grains of solid matter to the gallon, or $14\frac{1}{4}$ grains more than the Croton water, and hence was ill adapted to domestic uses. The population of the city had at this time become three times as large as in 1834, and the estimated cost of the proposed water-works was increased to $800,000. At the same time, plans were under consideration for a supply from the country streams, including an engine and stand-pipe at each mill-pond, forcing the water through iron pipes to a pump-well at Flatbush, from whence it was to pumped to a reservoir upon Prospect Hill. In 1851 the question was again taken up, and, reinforced by an appropriation to defray the expense of preliminary surveys, the committee obtained from Wm. J. McAlpin a report on the extent of the water supply and the mode of introducing it, as follows : "Dams were to be placed at Jamaica, Nostrand's, Springfield, and Simonson's Streams, and their waters carried by conduits into Baisley's Pond ; thence to conduct the water in a large conduit 9 miles to Flatbush, from whence it was to be pumped into an oval reservoir upon the summit of Prospect Hill of 60,000,000 gallons capacity. The minimum capacity of the works was to be 10,000,000 gallons per day, and the total cost, including a Cornish pumping-engine and 75 miles of pipe-distribution, $3,500,000." It was decided to submit this plan to the popular vote on January 27, 1852; but a few days previous to that date the election was deferred to further revise and mature the plans by means of more elaborate surveys.

In the meantime, a company known as the Williamsburgh Water Company had secured a charter, with the view of supplying water to Williamsburgh alone, and had purchased several of the ponds for that purpose.

In 1853 this company obtained an amendment to its charter, changing its name to the Long-Island Water

Company, increasing its capital from $500,000 to $3,000,-
000, and authorizing it to furnish a water supply to
Brooklyn also.

The election previously alluded to as deferred was
held on July 11, 1853, and resulted in a rejection by the
people of the proposed plan by a vote of 5,054 to 2,639.
In 1854 a new water committee was appointed, and a
new plan based upon former surveys was prepared for
submission to the people. This plan contemplated a
supply from all of the streams west of Parsonage Creek
at Hempstead; the overflow from dams to be erected
upon them to be conducted in an open canal to Baisley's Pond; from thence to flow through a closed conduit to a pump-well situated near Spring Creek, from
whence it was to be pumped into a reservior upon the
hill above. The works were to have a capacity of 20,-
000,000 gallons per diem when opened, and to be capable
of future increase to 40,000,000 gallons. The cost of
the whole was estimated at $4,500,000. This plan was
voted upon June 1, 1854, by the citizens, who again re-
jected the proffered supply by a vote of 9,105 to 6,402.
At about this time the well system again found its ad-
vocates, but was finally disposed of by the report of
J. S. Stoddard, C. E., which demonstrated that not more
than 1,000,000 gallons per diem could be obtained
through wells from a drainage area of 30 square miles.

In March, 1854, an offer was made by H. S. Welles &
Co , contractors, to contruct the works according to the
plans last voted upon for the sum of $4,175,000; but as
no arrangement could be made with the Long-Island
Water Company, which still held the ponds and reservoir site, this offer was not entertained. In 1855 another

company was chartered under the name of the Nassau Water Company, with power to construct works and to supply the city with water. Their plans included an open canal from Jamaica Pond, designed to intercept all of the smaller streams between the pond and the pumping-engines. On the 1st of January, 1855, the new City Charter, incorporating Williamsburgh and Bushwick with Brooklyn, came into effect, giving a fresh impetus to the water question. Under the authority conferred by the charter, negotiations were established between the Nassau Water Company and the city, the result of which was an arrangement whereby the city obtained a controlling power in the directorate of the Nassau Company by the subscription of $1,300,000, the company to conduct the construction of the Works in trust for the city pending the passage of a law enabling the latter to assume absolute proprietorship. A contract was at once entered into with H. S. Welles & Co. to construct works guaranteed to yield 20,000,000 gallons per day within two years from the date of contract. On the 31st of July, 1856, the commencement of the Works was formally inaugurated by breaking ground for the Ridgewood Reservior. From this time the work of construction rapidly progressed, until March of 1857, when a change in the character of the conduit east of Jamaica was found advisable, it having proved impossible, in the unstable sands, to construct or maintain an open canal that would continue reliable and efficient. Through delays attendant upon this change in the plan, operations upon this portion of the work were held in abeyance for more than a year. The remainder, however, was actively pushed to completion ; and on December

12, 1858, the long-looked-for supply was placed at the disposal of the citizens of Brooklyn. Thus, after a twenty-five years' agitation of the water question, its solution was ultimately attained, and the people have continued uninterruptedly to enjoy the convenience and protection of an agent which has contributed more than any other to the advancement of the growth and prosperity of the city.

TOPOGRAPHY AND GEOLOGY OF THE COUNTRY.

LONG ISLAND, upon the western extremity of which the city of Brooklyn is located, extends in an easterly direction about 114 miles, with a varying width of from 10 to 20 miles. Geologically considered, the Island is of very recent origin, having been formed almost wholly by glacial agency. If any portion of it were in existtence prior to the glacial epoch, it has been extensively modified and augmented by the eroding and transporting power of the vast field of ice which, extending from the far north, found its terminus in the warmer waters of the Atlantic. Upon the western end of the Island the origin of the materials of which it is composed is especially evident. The ground is full of worn masses of stone, with surfaces often deeply scored by obstacles over which they have been urged with resistless force during their transportation. Side by side, with bowlders of trap from the Palisades, dark trap and sienites from the Hudson Highlands, and the Taconic range, with occasional masses from the Lower Helderberg with their characteristic fossils, lie the gneiss and marble of West-

chester, the slaty shales of the Hudson group, the red
sandstones of New Jersey, together with many other
varieties of rock, the counterparts of which still abound
northward and northwestward of the Island. These,
mixed in extreme confusion, and packed with the
water-worn particles of their debris, comprise what is
known to geologists as the unmodified Drift. Throughout the entire length of Long Island runs an irregular
chain of hills of from 150 to 384 feet in height, the
northern spurs of which form the abrupt and diversified
North Shore ; while on the south the surface slopes
gently away toward the sea, terminating in a gravelly
plain, which skirts the shores of the bays with a varying width of from 5 to 15 miles. The material compos-
ing the ridge of hills is made up of the compact drift
already described. Bowlders are everywhere abundant,
and isolated beds of clay fill many of the depressions on
the elevated grounds, rendering them quite impervious
to water. This character is lost as the slope descends,
and layers of fine, uniform-grained sand, beds of pebbles
and gravel, and occasionally local deposits of clay in
thin strata, characterize the ground to great depths.
Through this porous material the waters flow toward
the ocean, bursting forth at various points in springs,
forming streams of singular clearness and purity. It is
from the larger of these streams that the water supply
of Brooklyn is derived.

THE DRAINAGE-BASIN.

THE area included between the ridge and the lowest limit at which the water of the streams is economized for the supply of Brooklyn comprises about 60 square miles, and constitutes the Drainage-Basin in which the rainfall is collected. The portion of this basin lying upon the immediate slope of the ridge does not yield a large, available quantity of water in proportion to its area ; for from the almost impenetrable nature of its soil, and the rapid slope of its surface, the greater part of the rain falling upon it .runs quickly off into the streams as storm-water, and is lost to the underground storage. It is from the great expanse of nearly-level, sandy plain that the larger portion of the supply is derived. This plain is composed of fine sand and gravel, very pervious, and at the same time very retentive of water, the fine interstices between the particles of sand affording room for the entrance of water, while at the same time retarding the velocity of its flow. The rainfalls of former years have been saturating this mass of sand ; and as each successive fall of rain sank into the ground faster than it could be carried off through the sand and into the bays, it followed that the level to which the saturation extended must continually rise until a sufficient head had been attained to force the water through the sand as fast as it was received. It is, by this process, that the water-roof, as it is called, has been formed—a roof, or upper surface of a water-bearing stratum of sand, sloping up uniformly, at a rate of 12 feet per mile, from the sea to the ridge. So accurately has the position of this underground water-surface been determined that it is possible to predicate, within a few feet, the exact depth at which water may be found anywhere within the bounds of the Drainage-Basin. The regimen of this water-bearing body of sand being once established, is evident that, whenever rain falls upon the surface, a certain portion of it goes to replace whatever may be wanting to fill the bed to the water-roof; the balance, unable to descend further, flows along the surface of the saturated sand, and finds its outlet in that class of springs which appear shortly after heavy rains and disappear during protracted droughts.

It further follows, from the extreme slowness with which the water finds its way through the water-bearing stratum, that the flow from the springs deriving their water from it is not perceptibly affected either by storm or drought. It is, in fact, the recorded opinion of engineers who have made this matter an especial study that it would require a succession of unprecentedly-dry seasons to materially affect the underground water-level or the flow therefrom. Such, then, is the nature of the source from which the Ridgewood water is derived. Flowing through a natural filter-bed, more perfectly adapted to its purpose than though it had been constructed by art, containing scarcely more than a trace of soluble mineral or organic matter, it is without a rival in the world for the purity and permanence of its supply.

THE PONDS

Traversing the surface of the Drainage-Basin from north
to south are several streams of various size, the largest
of which have been appropriated for water supply.
Commencing at Jamaica on the west, the first stream
met with is the supply of

JAMAICA POND.

THIS is probably the oldest artificial pond
on the Island. It receives the flow of two
streams, one of which rises about a quarter of
a mile south from the town of Jamaica and 4
miles from the bay, and the other has its source
near Little Plains, about the same distance from the
bay. After receiving each several small tributaries,
they unite near the head of what was formerly known
as Baisley's Mill-Pond. The valleys through which
these streams flow, though very wide are quite shallow,
not being over 15 feet below the adjoining uplands. A
small brook also enters the pond from the east, bringing
a small, additional supply. When first acquired by the
water authorities, the bed of the pond was filled with a
thick deposit of fine mud and vegetable matter; the
streams which supplied it flowed through a dense swamp
containing the vegetable debris of centuries, their waters
heavily laden with organic matter, and possessed at
times of a very disagreeable taste and odor. To remedy
these evils the pond was drained, and nearly 300,000
cubic yards of bog and muck removed, leaving a bed of

clean, white sand at the bottom. The channels of the streams were also widened and freed from vegetable accumulations, thus rendering the quality of the water all that could be desired.

During the excavation, remains of a mastodon were found, one tooth of which is cherished by the Long-Island Historical Society as an authentic memento of the oldest inhabitant. The old and insecure mill dam was removed, and a new dam built with an interior puddle-wall and paved slope, provided with an overfall 21 feet broad and sluiceway of stone-masonry, the latter delivering into a circular brick conduit of 42 inches interior diameter through which the water is conducted a distance of 2,937 feet to the main conduit. The drainage area from which this pond is supplied comprises 10.88 square miles, and its available storage capacity is 6,128,300 gallons. The area of its water-surface is 40 acres, and lies at a level of nearly 8 feet above tide. This pond can be relied upon for a daily supply of 3,135,-000 gallons, even in a season as dry as the dryest upon record. The distance of the pond from the pump-well along the line of the conduit and branch is about 5½ miles.

BROOKFIELD RESERVOIR.

THIS pond, situated at a distance of nearly 8 miles from the pump well, is supplied by a stream flowing through a comparatively narrow valley, and branching into three forks near its source, its total length, including tributaries, being about 4 miles. The valley is deeper than that of the Jamaica stream, and its rate of descent greater.

The power of the water was formerly economized at several points along its course, Simonson's Mill-Pond being the original name of what is now known as the Brookfield Reservoir. The dam is constructed—as is the case in all the ponds—with an interior wall of puddled clay, 7 feet wide at the base, and 9 feet high, the interior slopes paved and the outer sodded. The pond is rectangular in form, and has an area of water surface, when full, of 8¾ acres, standing at a level of 15½ feet above tide. It receives the supply from a water-shed of 5⅜ square miles, and delivers at a minimum a flow of nearly 2,000,000 gallons daily. The available storage capacity is 15,500,000 gallons. The Merrick plank-road crosses it near the centre of its length, dividing it into two parts, which communicate by a narrow raceway. The water reaches the main conduit through a circular branch conduit of brick 24 inches in diameter.

CLEAR-STREAM POND.

THE tributaries of this pond are short, their united length not exceeding one mile. The area drained by them barely covers 2½ square miles. The pond is constructed substantially as the others—of a somewhat triangular form, with a water surface of about one acre lying 11½ feet above tide. Its storage capacity is 800,000 gallons, and minimum flow about 750,000 gallons, per day. A branch conduit, 2 feet in diameter and 1,980 feet long, conveys its waters to the main conduit. It is distant from the pump-well about 8¾ miles.

VALLEY-STREAM POND.

BOUT one mile further to the east lies the Valley-Stream Pond, receiving the drainage of a narrow strip of land extending northerly to the ridge. The stream itself is nearly 4 miles long, without tributaries, and passes through two large mill-ponds in addition to that taken by the Water-Works and formerly known as Cornell's Pond. Its drainage area is 6¼ square miles, and its minimum flow 2,433,000 gallons, yielding a greater quantity of water in proportion to the area drained than any other of the streams. This is, perhaps, owing to the directness of its course and the steeper slope of the water-bed toward it. Its storage capacity is 18,800,000 gallons. The water surface covers an area of 17¼ acres at 12⅝ feet above tide. The branch-conduit of this pond is 30 inches in diameter and 2,103 feet in length.

ROCKVILLE POND.

LOWING into the pond is a stream which receives its supply from a strip of land parallel with and similar to that supplying Valley Stream, but separated from the latter by the valley of Foster's brook, which is not included in the Works. The collecting area is 8⅓ square miles, and the flow, at a minimum, 2,600,000 gallons. The stream is over 3 miles long, its course very direct and without important tributaries. In form the pond is long and irregular. Its water-surface lying 18 feet above tide, and covering an area of 8⅞ acres, represents a storage capacity of 3,128,000 gallons. The distance to the pump-well is 12¼ miles, the minimum flow 2,643,000 gallons, and the diameter and length of the branch conduit are respectively 2¼ and 1,872 feet.

HEMPSTEAD POND.

THE principal source of supply, and the most distant point from which the water is taken, is Hempstead Pond, situated 12.39 miles from the pump-well. It is fed by two streams originating a short distance from the town of Hempstead. These streams unite south of the town, and the subsequent course of the stream is marked by a succession of ponds of varying dimensions. The main pond at which the conduit terminates approaches the form of a square, with a rectangular projection at the point of entrance of the stream. Its water-surface embraces 23¼ acres, and stands, when full, 10¾ feet above tide. The drainage area extends over 25¾ square miles of the country, stretching northeasterly from the pond. Its storage capacity is 5,364,000 gallons, and minimum daily flow 7,800,000 gallons.

ADDITIONAL PONDS.

BESIDE those already enumerated, the city has gained possession of several other ponds, some of which form part of the streams flowing into the supply-ponds, while others, from circumstances of level and location, have not yet been found available. Willetts' Pond and Smart's

Pond lie upon Freeport Creek, 4¼ miles beyond the end of the conduit. This stream has a Summer flow of 5,250,000 gallons per diem; but the surroundings of the ponds are swampy, and would require extensive improvement to render the water desirable. One-Mile Pond, lying at the head-waters of the easterly branch of Jamaica Stream is small in capacity, and performs an unimportant part in the general scheme.

Quantity of Supply.

THE water flowing in the streams during the greater portion of the year would afford a copious supply for Brooklyn were the city double its present size; but seasons will come, often unexpectedly, when weeks and months of Summer drought succeed Springs of light rain-fall, and nothing is left to feed the ponds but the steady, yet slow, percolation through the sandy bed of the drainage-basin of the rains that have fallen long before. It is at such times as these, too, that the water finds its most lavish use in the city to refresh the thirst of its citizens, to cool its dusty streets, and to revive its drooping foliage. It is the chance of the occurrence of such a season —a chance which Brooklyn can ill afford to take, with the fate of Chicago and of Boston fresh in memory— that has recently led the Water Authorities to urge so strongly in favor of an increased supply. The least flow of the streams, guaged in a season of unusual drought, was found not to exceed 20,000,000 gallons per diem. The greatest flow has not been ascertained, and can only be inferred from the rain-fall. It has been de-

termined by able observers that only 42½ per cent. of
the total amount of rain falling in one year finds its
way into the ponds, the remainder being lost by evapor-
ation and in the supply of springs below the level of the
conduit. The rain-records of over 40 years place the
average rain-fall at 42¼ inches. At this rate the quantity of water brought down by the streams in one year
amounts to nearly 19,000,000,000 gallons! The consumption of water in the city in 1872 somewhat exceeded 8,000,000,000 gallons. The ponds have an united
storage capacity of only 57,000,000 gallons ; so that ten
thousand neglected millions flow annually over their
wiers. The daily consumption during the present year
has varied with the day of the week from 20,000,000 to
26.000,000 gallons, and upon one occasion has reached
33,000,000. During the Summer of 1872, owing to the
temporary interruption of the supply from some of the
ponds, pumping from streams below the level of the
conduit was resorted to until fortunate rains made good
the lacking portion of the supply. Early explorations
have determined the existence of 40,000,000 daily gallons of water in the streams between Hempstead and
the Connetquoit River, some 50 miles east from Brooklyn ; and many have favored the extension of the conduit in this direction, as was, in fact, contemplated at
the outset of the Works. Others, on the ground of
economy and compactness of the Works, have urged the
impounding of the surplus waters within the present
limits. The latter opinion has prevailed, but not without
a severe struggle. The contest of old was between
ponds and streams ; latterly it has been between streams
and reservoirs, no reflecting man in either case doubting

the necessity for more water, but each true to the last to his favorite mode of obtaining it. A new storage reservoir upon Parsonage Creek, above the Hempstead Pond, is now in progress, with a promise of completion in 1875, which will contain over 1,000,000,000 gallons of water. This will supply a deficiency of 10,000,000 gallons per day for over three months, and will set at rest for some time all fears of a water-famine.

The Conduit.

S before stated, the original design contemplated the construction of an open canal from Jamaica eastward, with branches to each of the ponds; but this crude and antiquated idea was abandoned as soon as the actual construction of a portion had demonstrated its inefficiency, and a substantial closed conduit of masonry was substituted throughout the entire length. Commencing at Hempstead, provision was made for carrying the 8,239,947 gallons derivable from Hempstead Pond, together with 20,000,000 gallons that might be subsequently collected from points further eastward; the conduit at this point having a width of 8 feet 2 inches and a slope or inclination of the bottom of 6¼ inches per mile. At the entrance of the branch-conduit from Rockville Pond the width is increased to 8 feet 8 inches; at the junction of Valley-Stream Branch to 9 feet 2 inches; at junction of Clear-Stream Branch, to 9 feet 4 inches; at junction of Brookfield Branch, to 9 feet 8 inches; while from Jamaica Pond to the pump-well the width is uniformly 10 feet, and the inclination of the bottom 6

inches per mile. This latter portion of the Conduit is
capable of delivering, with 32 inches' depth of water,
20,000,000, and, with 5 feet depth, 47,000,000 gallons of
water in 24 hours. The low level of the ponds and the
rate of slope made deep cutting necessary, especially on
approaching the pump-well. Through the whole distance the excavation was carried below the water-level
of the country, rendering the construction additionally
difficult. Portions of the bottom were found insecure,
and resort was here had to pile foundations, while in
other parts platforms of timber were used to afford a
firm bearing for the structure. The Conduit is support-
ed upon a bed of concrete 15 feet wide. On the lower
reach, between Jamaica and the pump-well, the sides
are of stone, 3 feet high, with an interior lining of
brick work. The bottom is an inverted arch of brick
4 inches thick, with a versine of 8 inches. The top is
a brick arch 12 inches in thickness. The height at the
centre is 8 feet 8 inches. The stone used is mostly
gneiss-rock from Greenwich, Conn. At convenient
points along the line, man-holes are added, either upon
the top or on the side of the Conduit, affording access to
its interior for inspection or for repairs. Not far from
the pump-well the water found was so considerable
that openings were left in the Conduit to admit it.
These openings, 30 in number, are estimated to be capable of yielding over 1,000,000 gallons per day. At
Spring Creek, as well as at a point about a mile beyond,
and at Jamaica, Valley Stream and Rockville Creeks,
are waste-wiers over which the surplus water escapes.
Here, a'so, are sluice-gates for draining the water from
the Conduit entirely, in the event of repairs becoming

necessary. The construction of the upper reach of the Conduit differs only in the omission of the inner lining of the side-walls, which at several points are made wholly of brick, and an increase in the thickness of the bottom to 8 inches. For almost the entire length the Conduit is covered with 4 feet of earth, sloping on the sides at the rate of 1½ horizontal to 1 perpendicular for the lower reach, and 2 horizontal to 1 perpendicular on the upper. Wherever the line has intercepted the courses of streams, inverted culverts have been built to carry the water beneath the bottom. At the pump-well the Conduit terminates in an arched basin 52½ feet long at right angles to its course, and connecting with the pump-well by four sluices. The total length of the Conduit is 12.39 miles, 7½ miles of which is 10 feet wide and 8 feet high, sufficiently large to drive a carriage through with ease.

The Pump-Well.

THIS structure, from which the water is drawn by the engines, is built of heavy granite masonry laid in courses with hydraulic mortar. The bottom, also of granite in radial courses, rests upon a bed of concrete, beneath which is a heavy platform of timber. The surface of the bottom is two feet below the bottom of the conduit. The interior of the Pump-Well is divided into compartments by cross-walls to admit of the examination and repair of either pump without interfering with the other.

THE PUMPING-ENGINES.

PON the plain near the foot of the hill, occupied by the Ridgewood Reservoir, and by the side of the pump-well, stand the Pumping-Engines, three in number, enclosed in a substantial; if not ornamental, structure of brick with brown-stone trimmings. The two chimneys for producing draught for the furnaces are 100 feet in height, and consist of two concentric shells of brick, with an annular air-space between them for 80 feet of the height, the diameter of the interior shell being five feet. Of the three engines now in operation one has been only recently erected ; the other two have been running since the completion of the Works. These Engines were considered at the time that they were built to be a modification of, and improvement on, the Cornish Engine, then much in vogue for pumping purposes. They differed, however, from the Cornish Engine radically, in that the steam was made to act directly on the column of water in the rising main, and that they were double acting, taking the steam at both sides of the piston and doing work at either end of the walking-beam. In the Cornish engine a ponderous mass of iron is raised by the steam, the descent of which, while the steam exhausts, forces up the water that has been drawn in beneath it. This class of engine is single acting, the return stroke being made by the gravitation of the pump-plunger. The only feature in common between the Ridgewood and Cornish Engines lies in the absence of the controlling power of the crank and fly-wheel.

The length of the stroke, within certain limits, depends entirely upon the pressure of the steam and the position of the valves, requiring the utmost watchfulness on the part of the attendants. The steam cylinder of Engine No. 1 was 7½ feet in diameter, and, of No. 2, 7 feet 1 inch, the length of stroke in each being 10 feet. The pumps are 3 feet in diameter and of the same stroke as the steam-piston. Upon the rising main of each pump are large air-chambers serving to regulate the flow of the water and to act as a cushion against any shock or sudden change in its velocity. The force-mains leading to the Reservoir are 36 inches in diameter and 3,450 feet long. The total lift is 164 feet. A sloping check-valve, with valves opening upward like trap-doors, prevents the return of the water through the force-main to the pump. The original capacity of these engines was about 15,000,000 gallons each in 24 hours, and their duty, when tested in 1860 and in 1862, was respectively 607,982 and 619,037 pounds of water raised one foot by the consumption of one pound of coal. These engines have been running almost continually from the time of their erection up to the completion in 1869 of the new engine known as No. 3. This third engine, after a year's trial, was found to operate so satisfactorily that in 1871 Engine No. 1 was dismantled, remodeled and converted into a crank-engine, with a fly-wheel 26 feet in diameter. The steam-cylinder was replaced by a cylinder 10 inches less in diameter, and the outer shell of the pump-cylinder dispensed with. These alterations resulted in the delivery of 3,388 pounds of water into the reservoir for every pound of coal burned in the furnaces. No. 2, which remains in its original condition, requires 1 pound

of coal for every 3,070 pounds of water thus delivered.
These two engines were built by Messrs. Woodruff &
Beach, of Hartford. The third and latest addition to
the engine power is Engine No. 3, built and erected in
1869 by Messrs. Hubbard & Whittaker, of Brooklyn. It
is of the class technically known as "Beam-Rotative
Engines." The fly-wheel is 26 feet in diameter and
weighs over 26 tons. The steam-cylinder is 7 feet 1
inch in diameter; the piston-rod extends through it and
acts as a plunger-rod for the pump beneath, the length
of stroke being 10 feet. At each stroke of the pump a
volume of water equal to 135½ cubic feet, or 1,059 gallons, is raised into the Reservoir. As the ordinary number of revolutions made per minute is 11, within that
short space of time nearly 12,000 gallons, or over 350
barrels, of water are poured into the Reservoir.

The Ridgewood Reservoir.

N the crest of the ridge above the Engine-
house is the Distributing Reservoir, covering with its slopes, embankments and grounds
a plot of 48½ acres. It is built in two divisions,
the area of which, when full, are 11.85 and
13.73 acres, or, together, 25½ acres. Its water surface,
when filled to a depth of 20 feet, is 170 feet above high
water. The embankments contain puddle-walls extending two feet higher than the level to which the Reservoir
is filled, and the bottom is covered entirely with a puddle
of clay and earth found upon the spot. The inner slopes
are paved with broken bowlder-stone upon a bed of
stone-chips and gravel. The water flows from the bell-

shaped mouths of the force-mains into a structure of heavy stone-masonry termed the influx-chamber, from which it flows over an apron and into each compartment of the Reservior. In traversing the distance of 1,200 feet, the velocity of the stream is quite lost, and ample time is afforded for the deposit of any sediment that may be held in suspension. At the effluent-chamber are two massive granite walls with 4 sluiceways in each, through which the water passes into the chamber and thence into the mains. Separating the effluent-chamber from the stopcock-chamber is a heavy wall of cut stone 6 feet in thickness, into which are set the mouth-pieces of 3 mains each of 36 inches diameter. Only two of these mains are in use at the present time, the admission of water to them being controlled by two large gates of stop-cocks in the stopcock chamber. Before entering the effluent-chamber the water passes through screens of copper wire, which arrest all floating matter of considerable size. Drain-pipes 12 inches in diameter, to drain the compartments for repairs, pass through the stopcock-chamber where their valves are placed. Of the two divisions, the Western is the larger, containing, when filled, to a depth of 20 feet, 86,651,382 gallons. The Eastern Division holds 74,439,062 gallons. The total capacity of the Reservoir is 161,090,444 gallons, which, at the present rate of consumption, is equal to about one week's supply.

PRINCIPAL MAINS.

FROM the northern wall of the efflux chamber passing through the stopcock-chamber, and curving thence westerly, extend the two principal mains through which the water descends by gravitation to the city. The main entering on the west side is the one first laid, and was the sole dependence of the city until 1867. It is 3 feet in diameter, and extends along the Cypress-Hills Plank-road to Cooper avenue, down Cooper avenue, through Broadway and Dekalb avenue to Vanderbilt avenue, where it is reduced in diameter to $2\frac{1}{2}$ feet. From thence it extends along Dekalb avenue, through Fulton avenue and Joralemon street to Clinton street, and along Clinton street to Hamilton avenue. A 30-inch branch-main proceeds from the corner of Dekalb avenue and Broadway, along Broadway to Union avenue; and another branch-main of the same calibre extends from the corner of Dakalb and Washington avenues, along Washington and Underhill avenues to the Prospect-Hill Enginehouse. Through the increased demand for water, this main soon became unequal to the supply; and in 1867 a second main was added. This main starts from the efflux-chamber parallel to the first, but soon bends to the south and passes along the Jamaica Turnpike and Atlantic avenue to Clinton street, where it is connected with the first main. Its diameter is 4 feet, and its total length $6\frac{3}{4}$ miles. From the ends of these large mains, and at various points in their length, 10 twenty inch mains extend, conveying the water to the extremities of the city, and acting as temporary feeders during the repair of any portion of the principal mains. The larger mains were laid in such a manner as to penetrate the then centres of population, their branches to reach toward and into the outlying neighborhoods. Their capacity is at present far in advance of the needs of the people;.but with the growth of the population they must fail, and an additional 5-feet main will eventually be laid on the north of the present mains and extended to the centre of the rapidly-growing Eastern District. The loss of head at present does not exceed ten feet during the hours of heaviest draught. The pipes of which the mains are composed are of cast-iron, varying in thickness with the pressure that they are called upon to sustain. A large number of them came from Glascow, Scotland; the remainder from foundries in New Jersey and Pennsylvania. At convenient points along the line of the mains gates are placed to control the quantity of water flowing through them, as well as to cut off the water for repairs. The gates are enclosed in brick chambers, entered through an iron man-hole set in the surface of the street. The spindles of all of the gates are geared for power, with bevel-gearing in the proportion of 3 or 4 to 1. The number of threads on the screws by which the valves are moved is usually either 4 or 5 per inch, so that it is necessary to turn the wrench 784 times with all the force that four men can conveniently exert to open or close a 48-inch gate. When the pressure is all upon one side of the valve, a much greater force than this is needed to raise it. From the number of the cross-connections between the mains there is little danger of the supply being cut off by a break in either, except in that portion of the 48-inch main between Nostrand avenue and the Reservoir, the failure of which would cause serious inconvenience upon the higher grades. The interior of the mains originally laid are now thickly covered with a coating of rust-nodules, or "tubercles," which resemble mushrooms in form, and are composed of oxide and carbonate-of-iron, with some clay. The inner surface of the pipe beneath these formations seldom appears corroded to any considerable extent ; and the question has been often raised whether the iron in the rust was originally a part of the water or of the pipe. The great disproportion in bulk between metallic iron and its hydrated oxide will amply account for the existence of a large amount of the latter without any readily-perceptible loss to the pipe. All of the mains and smaller pipes laid since 1862 have been coated, both inside and out, with a varnish of coal-tar and linseed-oil, into which each pipe is dipped, while hot, at the foundry. This coating affords an almost complete protection against the formation of tubercle.

MT.-PROSPECT ENGINE.

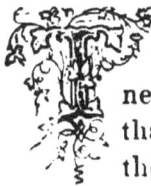

HERE are certain portions of the city in the vicinity of Prospect Park that lie so near to the level of the Ridgewood Reservoir that the water will not flow to them from thence, or, at best, will only furnish a limited supply during the night. For the accommodation of these the Mt.-Prospect Reservoir was built, and the Mt.-Prospect Engine erected, to supply it. The Engine is of the crank and fly-wheel pattern, with a steam-cylinder 24 inches in diameter and 4½ feet stroke, and two pumps each 20 inches in diameter and 3½ feet stroke. Being connected directly with the 30-inch branch-main, there is always a positive water-pressure of from 12 to 16 pounds per square inch upon the pump-pistons and a corresponding economy of power in operating them. There is an air-chamber upon both the induction and rising mains to render the flow of the water uniform. The ordinary speed of the Engine is 20 revolutions per minute, and the quantity of water raised at each stroke is nearly 112 gallons. From the pumps the water is conveyed through a force-main 20 inches in diameter and 2,052 feet in length to a height of 75 feet into the Reservoir.

Mt.-Prospect Reservoir.

ON a commanding eminence near the main entrance of Prospect Park is located Mt.-Prospect Reservoir. Its part in the system of water-distribution is to furnish, under a serviceable pressure, a proper supply of water to that portion of the city lying south of Atlantic avenue and east of Fifth avenue, which is too near to the level of the Ridgewood Reservoir to conveniently draw its supply from that source. The Reservoir grounds cover 11 acres, three fourths of which is occupied by the Reservoir and its appurtenances. The embankments are 20 feet wide at the top, and the slopes, both inside and out, are at the rate of 1½ horizontal to 1 perpendicular. Instead of the interior puddle-

wall of the Ridgewood Reservoir, the inner slopes, as
well as the bottom, are lined with a layer of puddle of
mixed clay and earth two feet in thickness. Upon the
face of this puddle an uniform bed of concrete, composed of hydraulic mortar and gravel, is laid 3 inches
thick, and upon this rests the brick slope-wall 8 inches
in thickness. The bottom has a slope of 6 inches
toward the south and is covered with a paving of
bricks on edge, with a fine mixture of cement poured
into their joints. In the northern embankment nearest
the Engine-house the influent-chamber is built, of cut-
granite masonry, 12 feet high, 6 feet deep, and 10 feet
long. The force-main from the pumps enters this chamber near the bottom. At a point some five feet above
the bottom a 30-inch pipe is placed, which passes down
under the inner slope. The water entering through
the first pipe fills the chamber until the level of the
30-inch pipe is reached, when it flows down the latter
and is delivered into the Reservoir beyond the foot of
the slope. The proper high-water mark is attained
where the water in the Reservoir is 20 feet deep, and
any surplus water that may be pumped into it after
that depth has been reached will flow out through a
12 inch overflow-pipe which starts from the influent-
chamber at that level. The high-water surface is 198
feet above tide, or 28 feet higher than that.of the
Ridgewood Reservoir. Upon the southern embankment stands the commodious Gate-House and Observa-
tory, the view from which, of Brooklyn, New York, and
their environs, is unsurpassed. On the south lies the
ocean, grand and still in the distance. On a clear day
one may count the breakers as they roll in upon the

sandy beaches of Rockaway and Coney Island. Toward the west spread the broad and fertile fields of Flatbush and the quiet beauties of Prospect Park. On the north is the woody summit of the Ridge, with East New York and Bushwick upon either side. Here may be seen how surely and insidiously the red and white hues of the city are stealing in upon the green farm-lands of the original settlers. On the southwest are the distant hills of Staten Island and New Jersey, and below them the broad expanse of the Bay, enlivened by the motions of the countless craft that ply its waters for pleasure or for gain. On the west the view extends over the three cities of Brooklyn, New York, and Jersey City—the homes of more than a million of people and the centre of the wealth and industry of the nation. The immense value of the cities exceeds conception, though all of their vast proportions are included in a glance.

The Gate-House is much more elaborately constructed than its office, in the system of water-distribution, would seem to require; but it was evidently deemed the least that could be done at a point where Nature had done so much. From the foot of the inner slope, at the rear of the Gate-House, a 30-inch main extends through a vault of stone-masonry and down the outer slope, carrying the water to the city. At the mouth of the main is a square chamber covered by a screen of copper wire upon the front and top to prevent the entrance of fish and floating impurities. A 12-inch drain-pipe follows the course of the main, which serves to draw off the last remnants of water when the Reservoir is emptied for cleansing or for repair. The gates controlling the flow through the main and drain-pipe are located in the vault. Connected with the drain-pipe is a vertical glass tube from which the height of the water in the Reservoir can be accurately read. The force-main through which the Reservoir is supplied has a 20-inch branch connecting with the effluent-main. The object of this arrangement is to enable the engine to deliver the water directly into the mains when for any reason it may be expedient to throw the Reservoir out of service. In such a case the 20-inch main leading up to the influent-chamber would act as a stand-pipe to regulate the flow. The capacity of the Reservoir is 20,000,000 gallons, and the daily consumption of water from it varies from 400,000 to 600,000 gallons.

PIPE-DISTRIBUTION.

THE interior diameters of the mains and pipes at present in use are 48 inches, 36 inches, 30 inches, 20 inches, 12 inches, 8 inches, 6 inches, and 4 inches. Originally the term " main " was applied to all pipes of 20 inches diameter and upward, and these were never allowed to be tapped for the purpose of private supply. Wherever water was required upon streets in which they were located, an additional pipe was laid. This was eventually found to result in a needless multiplication of pipes, and the 20-inch mains were permitted to be tapped in the same manner as the smaller pipes. At the outset, the ruling length of the pieces of which the mains and pipes are composed was 9 feet. They were then cast horizontally, or nearly so ; and with a greater length, the sagging of the core, or mould of the interior, would have resulted in an inequality in thickness between the top and bottom. This mode of casting has since been changed, all of the pipes used at present being cast vertically and 12 feet in length. They are connected by inserting the spigot, or straight end, of one pipe into the hub, or bell-shaped mouth, of another, driving several strands of hemp into the annular space between the spigot and hub, pouring from 2 to 2½ inches of melted lead into the space, and finishing the joint by driving the lead compactly home with proper tools. This kind of joint admits of considerable movement of the pipe before it shows signs of leaking, and is readily made water-tight again by a few blows of the hammer. The pipes used are divided into two classes, according to their thickness, and designated as A-pipe and B-pipe by having these letters cast upon them. The A-pipes are used upon grades higher than 50 feet above tide, the B-pipe being used upon the lower grades. The pipes are cast by contract, at various foundries in New Jersey and Pennsylvania, among which are those at Camden, Florence, Conshohocken and Phillipsburgh. The Water Department invariably stations a resident inspector at the foundry where the pipes are being cast. It is his duty to look to the quality of the iron or ore used in the furnaces, and to protest against the introduction of improper material. When the pipes are cast, they are cleaned from the adhering portions of mould and core, and submitted to a preliminary hammer-test by the inspector. This often results in the discovery of cracks from unequal shrinkage or from careless handling, porous cavities caused by the cooling of one portion in advance of another, masses of sand

broken loose from the core and become imbedded in the
iron, together with many other objectionable defects
sufficient to warrant him in rejecting the casting without further trial. The pipes that are found apparently
free from these defects are heated and immersed in a
protective varnish and placed in the proving-press,
where they are subjected to an interior hydraulic pressure of from 250 to 300 pounds per square inch. While
under this pressure, they receive several sharp blows of
a hammer. If this test be satisfactory, they are weighed, marked. and forwarded to the Pipe-Yard at South
Brooklyn, where they are again weighed and sent out,
as occasion demands, to become a part of the general
distribution. Each pipe has its number, class-letter,
date, and name of maker cast upon it, beside the weight
marked in white paint. The average weight per foot of
the several sizes of pipe is as follows : 4 inches, 24 lbs. ;
6 inches, 37 lbs. ; 8 inches, 49 lbs. ; 12 inches, 76 lbs. ; 20
inches, 180 lbs. ; 30 inches, 340 lbs. ; 36 inches, 410 lbs. ;
48 inches, 712 lbs. The total weight of the iron pipe now
lying in the streets exceeds 50,000 tons. In addition to the
cast-iron pipe, there is about 2 miles of cement-lined
wrought-iron pipe in use. The limited amount of this
pipe that was laid was due to the prevailing distrust of
its durability at that time. No leakage, however, in the
existing 2 miles of pipe has been discovered during the
past three years. At points where it is desirable to connect the pipes, special castings in the form of a cross or
letter T are inserted ; and also where fire-hydrants are
located, branches for that purpose are put in. Formerly
the branch-pipes leading to the fire-hydrants were only
4 inches in diameter ; at present they are increased to

6 inches, which admits of the use of several hoze-connections on the same hydrant, and more than doubles the supply of water. Seven different patterns of fire-hydrant have been used upon the Works, but none have proved so reliable as the original, Coffin hydrant, which, notwithstanding its unsightly wooden box, has given less cause of complaint, and has required less repair than any other form that has yet been tried. The total number of fire-hydrants now in position is 2,208. There are also 30 surface-hydrants rising directly from the mains and covered with an iron manhole-casting set in the surface of the street. These hydrants have a movable head containing the hose-outlets, which is kept until wanted at some convenient locality near to the hydrant. Drinking-hydrants, originally designed on the Holly-Tree principle, to rescue the pedestrian from the allurements of the bar, have latterly been erected for the supply of people whose means do not permit them to introduce the water into their houses. They are a perpetual joy to amateur hydraulicians, and at such times a terror to the passer-by. Being chronically out of order, they are a prolific source of expense and waste to the Water Department. The present number upon the Works is 850. Along the river-front, and at several points on the line of the principal mains, are blow-offs, or outlets, for the purpose of drawing off the water for repairs to the pipes or to free them from sediment. The former discharge directly into the river—the latter into the sewers or into basins especially constructed for that purpose. The leading design in planning the pipe-distribution has been to introduce into the various sections of the city one, or if practicable

two, mains from different directions, and preferably
from different sources, each of which shall prove ade-
quate to the supply of that section when it shall have
been solidly built. With this arrangement, in the
event of an accident to either main, the other can still
be relied upon. The connections with the principal
mains are infrequent, lateral branches occurring only at
long intervals, and provided with gates at such points,
both upon the branch and upon the main. In the case
of the 20-inch pipes, the ramifications are increased, frequently connecting with 6-inch pipes, which are always
provided with a gate near the branch. This latter size
is that principally laid for the general supply, while 8
and 12-inch pipes are interspersed at proper intervals,
not that the streets in which they are laid require more
water than others, but to facilitate the circulation and
to serve as feeders to convey the water around any par-
ticular portion of the main that may be temporarily
thrown out of use. The gates of the smaller pipes are
set in wooden boxes with iron covers, usually on the
·building line of the street and six feet from the curb.
The number of gates of all sizes now in use is 1,793.
The pipes are all laid at a depth of 4 feet below the
surface of the pavement, and where practicable, upon
the north or east side of the street, six feet from the
curb. Some few exceptions to this occur in unusually
wide streets and in those paved with an expensive
pavement, in which case a pipe is laid under the side-
walk on either side of the street. Each separate supply
is obtained by drilling a hole into the pipe and inserting
a brass tap. This tap is perfectly smooth upon its driv-
ing-point, and is not screwed, but simply driven in, re-

taining its place by friction. It has a movable plug similar to that of an ordinary faucet, and a coupling to connect it with the lead pipe, four feet of which is always laid next to the tap, although iron pipe may be used to carry the water into the premises. The waterway of the taps used for the supply of private dwellings is three eighths of an inch in diameter. Factories and public buildings are supplied from half-inch and five-eighths-inch taps, or from several of these connected with a single pipe, while large manufacturing establishments, as sugar-houses, breweries, etc., are supplied through four-inch pipes. Many of these have also an independent line of four-inch pipe, carrying the water throughout the entire buildings, with hose-attachments in every room, kept in readiness for immediate service in case of fire. House-services are of lead, tin-lined lead, iron, galvanized iron, and cement-lined pipe. None of these are without their disadvantages. The lead pipe poisons the water, the tin-lined pipe is difficult to connect without melting off the lining, the iron pipe rapidly fills with rust, and the galvanized iron is even worse than the lead, in that wherever the zinc flakes off, or wherever a brass cock is placed, a galvanic action ensues at the expense of the zinc, contaminating the water with a metalic salt scarcely less detrimental to health than are the salts of lead. The cement-lined pipe has not been used sufficiently to test its qualities.

The original pipe-distribution comprised 10 miles of mains and 110 miles of distribution-pipes. This amount has been annually increased until at the present time the total length is 16¼ miles of mains and 276¼ miles of pipe, or 292¼ miles in all. Four submerged lines of pipe have been laid, two of which are still in use. One of these—a 12-inch pipe—supplies that portion of the city which lies along the shore of Gowanus Bay, and crosses the Gowanus Canal at the Penny Bridge ; the other—a 6-inch pipe—lies in the bed of the Wallabout Channel, and carries the water to the Ordnance Dock of the Navy-Yard. Both were laid with a movable ball and socket-joint—an invention of the Water Purveyor, Mr. J. H. Rhodes—and have both proved water-tight and reliable.

PIPE-YARDS.

NEAR the mouth of the Gowanus Canal is the Pipe-Yard, at which the pipes and other appurtenances are received, inspected and kept until required for use. On Portland avenue, between Park and Myrtle, is the Repair-Yard, where all of the material and implements necessary to remedy any defects in the distribution are stored. Here all of the water-meters are tested and repaired, and all the perishable portions of the hydrants, gates, etc., kept constantly on hand. The number of tools, patterns, fittings, and miscellaneous supplies is immense, but all are arranged with such admirable system and order that each is readily found as occasion requires. The office of the Repair-Yard is connected by telegraph with the City-Hall, the Central Police-Station, the Reservoir, and the Engine-House. This Yard is the headquarters of a corps of able and efficient men, who stand ready at all hours of the day or night to attend to any accident to the distribution. Water is popularly supposed to be one of the most innocent of substances; but when bursting from its confinement, with a pressure equal to that ordinarily existing in steam-boilers, its capabilities for mischief are manifest. The veterans of the Pipe-Yard tell of many scenes of excitement, if not of danger, at their midnight "leaks in main." A smaller establishment on North First street affords a rendezvous for a similar corps of men, though fewer in number, who keep the distribution of the Eastern District in a thorough state of repair.

Consumption of Water.

WHEN the water was first generally introduced, the average daily consumption barely exceeded 4,500,000 gallons, or at the rate of 25 gallons per day for each inhabitant. This consumption has continually increased, not alone in its total amount, but also in the rate proportioned to the population, which has now reached 46 gallons per capita This increase in the rate per head is due in a measure to the increased facilities for disposing of waste-water, which the extension of sewers affords, and to a growing recklessness upon the part of the people in the use of an element apparently so abundant. It is also in part due to the growth of manufacturing industries The average daily demand for water during the year 1872 was 22,700,000 gallons. The demand reaches its maximum in the coldest part of the Winter and in the heat of Summer. The greatest use occurs on Monday and Friday; on Sunday it is from one quarter to one third less. At the different hours of

the day the consumption varies widely, commencing an hour before daylight, increasing steadily until, in the Summer months, between 8 and 9 A. M., about one tenth of the day's supply is drawn. It decreases abruptly with the ringing of the bells at noon, grows again to nearly its forenoon proportions at 3 P. M., and gradually diminishes, with a slight quickening between 6 and 7 P. M., until the lowest rate of the day is reached, between the hours of 12 and 1 A. M. This lowest rate is never less than 200,000 gallons per hour, and frequently exceeds 500,000 gallons.

The use of water for manufacturing purposes is large and continually increasing. The water consumed by the fourteen sugar-houses alone averages 300,000 gallons per day, and the numerous breweries, steam-engines, and steam-vessels add largely to the demand.

The use of the water as a motive power has not thus far met with the extended patronage that its convenience merits. The principal use that is made of the force of the water at present is in the blowing of church-organs. The hydraulic-engine is usually located in the cellar, and a valve within the reach of the organist serves as a more expeditious means of awakening the moving power than in the case of muscular aeration. The cost of producing a hymn to the tune of Old Hundred upon the organ of Plymouth Church—at the present price of water, 2 cents per 100 gallons—is 14½ cents; but the same familiar strain, in its puritanic simplicity, can be extracted from less pretentious instruments on lower grades for about 10 cents. The higher a church is— that is, the higher above tide—the more expensive is its music, the water having less pressure; more of it is required to furnish the necessary air to the organ. The cost of obtaining power for manufacturing purposes from the water is, within 50 feet of the tide, not far from 50 cents per horse-power per hour. The pressure in the business part of the city is ample, varying from 40 to 65 pounds per square inch, according to grade. The use of these water-motors is attended with great economy where the demand for power is small and quite intermittent; but for a steady work steam is much the cheaper. The use of water for manufacturing and for business purposes is, to a considerable extent, controlled by meters. There are at present seven different kinds of water-meters in operation upon the Works. Three of these are positive measures, two of which are operated by pistons and reciprocating-valves, and one by diaphragm and valves; the remaining three are indicators, two consisting of propeller-wheels actuated by the water flowing by them, and one on the paddle-wheel principle urged by the impact of water against the blades. The piston-meters give ordinarily the most equitable measurement, the indicators invariably favoring the consumer, as on very slow streams they fail to record. The number of meters at present in use is 649.

QUALITY OF THE WATER.

THE quality of the Ridgewood water compares very favorably with that of any other water-supply in the world. It has none of the hardness characteristic of the Croton, which flows over a primitive foundation, or of the supplies of Troy and Newburgh, derived from a

country where calcareous and aluminous minerals
abound. This quality renders it more economical for
domestic uses in the saving of soap and for manufacturing purposes, in that it does not cause any considerable incrustation in steam-boilers. Its inorganic impurities are of a sedimentary character and settle to the bottom of the boiler in the form of fine mud instead of
forming a scale or incrustation, as would be the case
were the mineral impurities held to a greater extent in
solution. It does not injuriously affect the health of
the occasional visitor, as is frequently the case with the
waters of neighboring cities. The soil through which
it flows is composed of almost entirely insoluble material, and the complete aeration that it receives at the
broad surfaces of the ponds and reservoirs causes the
precipitation of the greater part of the mineral impurities held in solution. The soluble inorganic matter
consists of the carbonates of magnesia and of lime (the
latter present to a greater extent in Jamaica stream
than in any of the others), chlorides of magnesium,
calcium and sodium, sulphates of lime and magnesia,
and oxide of iron, the weight of the whole not exceeding 2¼ grains per gallon, or in about the proportion of
one ounce in six barrels of water. For the sake of
comparison, the analyses of the waters used in several
cities are added, the numbers indicating the ˙grains of
solid matter per gallon : Brooklyn, 2.64; New York, 6.65 ;
Philadelphia, 4.26 ; Boston, 3.57 ; Albany, 4.72 ; London,
28 ; Paris, 9.86. This exceeding purity of the water is
not, however, without its objections, since it fails to at-
tach that protective coating to the pipes that is deposit-
ed by more alkaline waters, but, on the contrary, it

actively attacks all lead and iron with which it comes
in contact. Small iron pipes are often closed in a few
years with rust, while in pipes of lead and galvanized
iron white, glistening particles of oxycarbonate of lead
or zinc are formed, which have proved in numerous
cases detrimental to the health of consumers. Perhaps
the most objectionable impurity of the Ridgewood
water is its organic matter, which, happily, is not present in any considerable quantity. It is derived from
the decay of the luxurious vegetation of the country
through which the water flows, in part, and in part
from the imperfect cleansing of the beds of the streams,
as well as from the multitudes of fish that infest them.
Wherever the waters are impounded they swarm with
fish, the most numerous of which are the yellow perch
and eel; but the trout roach, shiner, pickerel, white
perch, and black bass are all well represented. These
fish are all animal feeders, and contribute nothing
toward the suppression of vegetable growth. Complaints of a disagreeable taste and smell of the water are
rare and local. When they occur, the difficulty is
speedily removed by opening a fire-hydrant in the
vicinity for a short time. On such occasions, the water
flowing from the hydrant yields, if filtered through a
cloth, an immense number of fine, greenish-brown con-
fervoid filaments, rarely branching, but much interlaced
and often curiously beaded with short nodes alternately
lighter and darker in color. Complaints of the mudiness
of the water are more frequent, and these arise from a
disturbance of the rust and sediment in the pipes by an
unusual velocity of the water. This difficulty it is im-
possible to remedy, as the opening of hydrants only

increases the disturbance. ˙ It occurs most frequently on
Monday mornings in the Summer, when, in addition to
the great demand for water for washing, the street-
sprinklers are also at work. A serious inconvenience to
the consumers arises from the presence of eels, which
find their way through the screens while they are still
very small, and, passing down the mains and into the
smaller pipes, collect in great numbers upon the lower
grades. They grow to great size, often exceeding three
· feet in length, and find a plentiful subsistence on the
swarms of fresh-water shrimp with which the pipes
abound. On the approach of Winter, usually at the
first decided fall of temperature, the trouble begins
from their entrance into the taps, and as many as
twenty consumers have had their water-supply thus
summarily cut off in a single day. The obstruction
usually defies all attempts at remedy short of digging
out the tap and extracting the unfortunate explorer
with a cork-screw. The eels taken from the pipes are
fat, well conditioned, and apparently not inconvenienced
by the pressure of from 50 to 75 pounds per square inch
that they have sustained. It is a singular fact that
these eels are almost invariably caught by the tail.
Where one eel has been taken from a tap the chances
are in favor of obtaining several more at the same tap
in quick succession. A contrivance has recently been
applied to the taps, which will prevent all trouble with
the new taps in the future ; but as there remain some
40,000 taps unguarded, the people of Brooklyn will be
obliged for a long time to carry on an expensive fishery.
Frogs of over half a pound in weight, and trout from
four to six inches long, have occasionally been thrown

out through the fire-hydrants. The eyes of the fish
taken from the pipes are frequently covered by a nearly-
opaque pellicle through which the pupil is barely to be
distinguished.

Cost of the Water.

THE cost of the Works, under the original
contract, was $4,625,000, and the expense
of extending and perfecting them during the
subsequent years has more than equaled this
amount. The total cost at present is $9,500,000,
and their value, or the cost of constructing them, at the
existing rates for land, labor, and material, has been estimated at $15,000,000. The amounts necessary for con-
struction and extension have been raised by the issue of
bonds, the interest on which, together with the cost of
maintenance and an annual deposit of $50,000 as a sink-
ing-fund for their retirement as they become due, has
been paid from the revenue derived from the water-
rates. Up to the year 1871, a large and increasing deficiency existed, which was provided for by general
taxation. In that year the water-rates were raised to an
extent that resulted in an increase of the revenue of 30
per cent., leaving a surplus as large as the former deficiency. The rates have since been so adjusted as to
cover as nearly as possible the necessary expenses of interest, deposit, and maintenance, and the surplus, if any,
is by recent legislation made a part of a special sinking-
fund. The water-revenue is collected in two forms—as
regular and as extra rates. The regular rates are cal-
culated upon the front dimensions of the building before

which the distributing-pipe passes, or, in the case of vacant property, upon the frontage and assessed valuation. Each building is entitled to the use of one bath, one water-closet, and the necessary sinks and basins for the domestic uses of 12 occupants without charge further than the payment of the regular rates ; all water-fixtures and occupants in excess of these are subject to additional charge. The regular rates accrue whether the water is introduced into the premises or not, and become a lien upon the property in the same manner as the city taxes. The extra rates are collected for the use of water for other than domestic purposes, and their payment is enforced by the cutting off of the supply. The rates are established upon the basis of a charge of 3 cents per 100 gallons of water used for domestic purposes, 2¼ cents for a like amount of building, stabling, fountain, and tavern uses, and 2 cents per 100 gallons for business and manufacturing purposes, steam-engines, etc. For the latter uses, the quantity taken is accurately determined by meters ; but in the other cases, as there is no means of limiting the quantity consumed by each, the rate is far from being uniform or equitable. During the year 1872, the cost of pumping alone was one eighth of one cent per 100 gallons, and the total cost of delivering 8,250,000,000 gallons was at, very nearly, the rate of 1 cent per 100 gallons. Not far from 35,000,000 gallons were used for fire and sewer purposes, and about 3,000,-000,000 gallons were paid for at the rates above, leaving over 5,000,000 gallons furnished gratuitously. It is evident that an enormous waste of water is allowed by many negligent members of our community ; in fact, the average consumer, who pays but $10 a year for the use of water, may, by leaving his faucet open, waste in one year $300 worth of water, at the established rates, or water that it has cost over $20 to pump into the reservoir. A general system of supply by measurement through meters would, no doubt, curb the extravagant waste, but might, on the other hand, induce an economy of water that would be prejudicial to the public health and cleanliness, and interfere with the proper discharge of the more solid portions of the sewage. A reduction in the size of the taps, or the use of meters designed to discontinue the flow when the amount of water paid for by the rate fixed upon the premises shall have been drawn, would seem a medium course and one promising favorable results.

It is believed that the foregoing pages have been sufficiently explicit to convey to the interested citizen or to the inquiring observer from abroad an intelligent idea of the construction and operation of the Water-Works of Brooklyn.